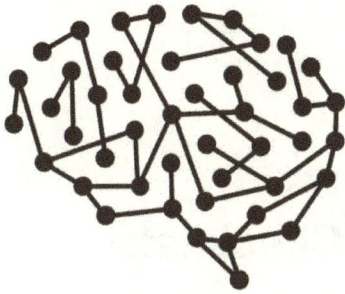

KISP

Prof. Kunow + Partner

Annette Kunow

7 Buoyant Ckecklists

for Your Project

Goal-oriented and efficient project management with correct body language and comprehensive communication

COPYRIGHT © 2019

NAME: Annette Kunow

ADRESSE: Baumhofstr. 39 d, 44799 Bochum

Web: www.kisp.de

E-Mail: info@kisp.de

Tel: 02349730006

Illustration: Annette Kunow

ISBN Nummer: 978-3-96695-004-6

PREAMBLE

Due to the very lean hierarchies in today's companies, the targeted control of even smaller projects and skilful human management become the yardstick of project success. Some of these factors and requirements for the project manager are the subject of this book.

Checklists and empty forms are used to provide suggestions for submitting appropriate documents for your own project.

The target group are practitioners in industry and students from all disciplines who want to work as project managers in these areas.

Bochum, spring 2018

Prof. Dr.-Ing. Annette Kunow

Here you can book a free Strategy Session or write me if you like this book and have suggestions or questions.

Click here for free bonus material on the book.

Also visit my blog „Selbstführung & Produktivität in German". I'll help you get better results.

INHALTSVERZEICHNIS

1 EINFÜHRUNG

The first step is halfway.

In order to plan correctly, you need the right hand tools or the right equipment.

I'm an advocate of checklists. Therefore I show you here my 7 most important checklists for planning and successful execution of a project.

And you get a comprehensive description of the individual steps to be able to implement it well.

2 PROCEDURE FOR PROJECT DEFINITION

To make a project successful, it needs a good project definition (https://www.kisp.de/erfolgreiches-projekt/). The project is considered from all sides if possible.

The best thing to do here is to proceed step by step.

- o **Study, discuss and analyse the topic**
  It is very important that the team has enough time to study, discuss and analyse the project, as this gives everyone a clear idea of the problem. It may be necessary to investigate the approaches in other projects or to what extent other experience patterns could contribute to project planning.
  The purpose of this activity is to clarify that the right problem or opportunity is being considered.

- o **Preliminary project definition**
  If the project manager thinks he has the situation under control, he creates a preliminary project definition. This will of course be continuously revised with the arrival of new information and new experiences.

- List needs and wishes

  First, a list of the absolute necessities (need to have) that must be realized in the end result of the successfully completed project is drawn up. Then-after a second list can list the additional wishes (nice to have), whose realization is not necessarily required for the successful completion of the project, but would increase it.

- Set final destination

  Either the final goal for the project must be determined with the help of the project definition itself or the final goal is determined by the client from the outset. After that the feasibility is checked with the help of the project definition.

- Develop alternative strategies

  Alternative strategies are now being developed for the solution strategy, which could also lead to the goal. To find these alternatives, it is advisable to use creativity methods such as brainstorming sessions or creating mind maps to exploit the full creative potential of the project team.

- Evaluate alternatives

  The next step is to evaluate the alternative strategies found and worked out in this way. It must be ensured that the selected evaluation criteria, for example a marketing study or a risk analysis, are realistic and reflect the final objective.

- o **Select procedure**

  This evaluation now allows a more objective selection of an approach that best corresponds to the project definition and the final goal.

This process may have to be repeated several times before the desired result is achieved.

Today, Design Thinking is often used for project definition in order to develop the project from all sides.

Design Thinking refers to creative strategies originally used by designers during the design process. Today it is also used as an approach to solving problems in other contexts, for example in agile project management.

3 COST CATEGORIES

- o **Labour costs**

  The wages and salaries of all workers and employees working directly on the project for the
  time spent on it

- o **Overheads**

  The employer's contributions and fringe benefits
  for project employees (generally a percentage
  of total labour costs).

- o **Material costs**

  The cost of all items used in the project, such as
  timber, cement, cables.

- o **Equipment costs**

  The cost of tools, equipment, office supplies,
  etc. required for the project; if a facility is used
  for the duration of the project, the costs may
  have to be allocated over the duration.

- o **Equipment rental**

  Rental costs for large equipment, such as scaf-
  folding materials, compressors and cranes

- ○ **Additional costs**

  The cost of management and support services such as purchasing, accounting, secretariat, etc., for the time devoted to the project (generally a percentage of the total project cost).

- ○ **Profit**

  The profit on successful completion of the project (generally a percentage of the total project costs).

Once the individual cost categories have been determined and the project is divided into individual steps, the budget for the entire project is determined.

The costs of the individual steps, which are incurred, for example, by a company contractor, are generally much easier to estimate than their own, because today most contracts are awarded at a fixed price. These costs are then composed of the agreed price, the fixed price, and the costs for selecting and awarding the sub-project.

4 DEFINITION OF GOALS

In order to be able to develop further, you must constantly question your value systems.

Every one of us has a flash of genius at one time or another. But the opportunities and conditions for turning it into reality are not always equally distributed. Many of chances for fame and recognition, an idea for a new book, a new product or a marketing strategy, could not be exploited simply because the right tool was not available to put the idea into practice beyond the critical phase of planning.

Success means something more than just having a great idea. Rather, it stands or falls with the ability to transform these thoughts into tangible goals on the basis of which project plans can be worked out, milestones set and deadlines reached. However, since goals can rarely be consistently pursued and achieved without considering the environment, flexibility and the ability to adapt to the permanent change that a dynamic work environment is subject to are also required.

That is why it is necessary to define a goal precisely. The method of neurolinguistic programming (NLP), a descriptive, explanatory and teaching model for communication, is process-oriented.

It is always possible to work without disclosing the objective.

This is particularly helpful when working in conflict situations in companies. Only the observation of his reactions and the questioning with a certain questioning technique can show the experienced project manager how he can influence the behaviour of the employee without hurting his feelings.

In the "content-free" survey, the questioner does not know the goal of the interviewee, since he does not want or cannot reveal his goal. However, the interviewer must ensure that the interviewee has a sufficiently precise idea of the chosen target image.

These questions will help you to describe your goals in detail.

- o   What do you want to achieve?

- o   What is your goal?

- o   How exactly do you want to behave towards whom and when?

- o   What do you want to be able to do when you reach your goal?

It is important to pay attention to positive formulations, i.e. no negations.

Comparisons in the target shrinkage are also misleading. Often they are even open comparisons, as they are often used in advertising, for example "Get richer!

Defining of goals is essential

But without value definition, there is no definition of goals that is successful.

If you imagine that you are what you think you are, then it is not surprising that you yourself bring about what you like. So this also applies to your goals.

At the beginning of your life, you take over your value system from your parents and your immediate environment. These are your beliefs that will become your life guidelines.

In puberty you question this for the first time and gradually develop your own value system.

Here you can ask the following questions.

- o What is important in your life?

- o What do you want to achieve?

- o What is your goal?

- How exactly do you want to behave towards whom and when?

- What do you want to be able to do when you reach your goal?

Thinking - Beliefs   I'm what I think

$\downarrow$

Coincidence

Effect   It falls to me.

$\downarrow$

Success follows

Beliefs

Questions such as: "... where exactly, how exactly and to whom exactly...?" achieve a good analysis concerning the context, for example at which point in time or in which framework the target behaviour makes sense.

To get clarity for the goal, the desired state is described and addressed with all five senses (visual = seeing, auditory = hearing, kinesthetic = feeling, olfactory = smelling, gustatory = tasting).

For example, a car salesman sells a car best with the following words: "Take a look at the great shape of the car body (V). And then the sound of the engine (A). And are you not sitting extremely comfortably in the sports seats (K)? And smell the new leather (O)!" But you can't taste the car (G)!

However, if the content of a target is published, the questioner does not need to understand everything literally, because words are only metaphors whose real meaning is often not even quite clear to the person seeking the goal at the beginning of the process.

In addition, it is sometimes the very target of a discussion to initiate a search process to clarify or translate these metaphors.

If the goal changes during the discussion, this usually goes hand in hand with an increase in the target seeker's satisfaction. This becomes visible in his body expression, for example a symmetrical posture and a satisfied visual expression. The discussion leader promotes this process by helping the goal seeker to formulate his goal ever more precisely. In this way, the statements become more consistent and the goal gains contours. The goal seeker looks reconciled and symmetrical. He is now aware of his abilities.

Concrete examples for definition of goals (hidden and open) can be found in detail in the book "Project Management & Business Coaching" (https://www.kisp.de/buchshop/projektmanagement-business-coaching/)

S.M.A.R.T. Goals

In project management, the term S.M.A.R.T. has developed for clear goals with S.M.A.R.T. criteria. Every team member must be able to accept the goal.

Definition

The term S.M.A.R.T. comes from the English-speaking world and is used in project management as a criterion for the clear definition of goals within the framework of a target agreement. S.M.A.R.T. is an acronym for "Specific Measurable Accepted Realistic Timely".

It can be interpreted like this: An objective is S.M.A.R.T. only if it meets these five conditions.

- o **S Specific:** Goals must be clearly defined. Not vaguely, but as precisely as possible.

- o **M Measurable:** Goals must be measurable (measurability criteria).

- o **A Executable** (achievable)**:** The recipients must accept the goals. But they must also be appropriate, attractive or demanding.

- o **R Realistic:** Goals must be possible.

- o **T Schedulable:** Each goal requires a clear deadline by which it must be reached.

Here you can book a <u>free Strategy Session</u> or <u>write me</u> if you like this book and have suggestions or questions.

Click here for free <u>bonus material</u> on the book.

Also visit my blog „<u>Selbstführung & Produktivität in German</u>". I'll help you get better results.

5 TAKE CORRECTIVE ACTIONS

As the project progresses and performance is monitored, there will always be times when the project falls behind schedule. This generally always requires corrective action, but the project manager should also be careful not to intervene too quickly. Some defects correct themselves.

Sometimes the project will fall behind schedule, sometimes it will progress faster than expected, but eventually it will be "on schedule".

It would be unrealistic to always expect steady progress according to plan.

If the project actually begins to follow the plan, there are several ways to deal with it.

- **Renegotiate**
  Discuss with the customer the possibility of extending the deadline or increasing the budget.

- **Catch up**
  If time was lost in connected single steps, the budget and the schedule for the remaining work must be controlled. Perhaps savings are possi-

ble there in order to meet the cost and time targets after all.

o **Limitation of the project scope**
Unimportant project steps (wish list) can be cancelled in order to reduce costs and save time.

o **Increase resources**
More employees and machines can be deployed for the project to meet an important deadline. The costs for this must be compared with the importance or the benefits.

o **Search for replacement**
If something is not available or more expensive than expected, replacing it with comparable materials or equipment can save money or make the work possible.

o **Alternative sources**
If a supplier cannot deliver within the specified time or cost frame, a replacement for its scope of supply should be sought immediately. However, the alternative offered by the supplier can also be accepted once before it is replaced.

o **Partial deliveries**
Sometimes a supplier with partial deliveries can

help to keep a project on schedule and deliver the rest later.

o **Bonuses**
Going beyond the limits of the original contract or offering a bonus or similar incentives may still allow timely delivery.

o **Demand contract compliance**
Sometimes it is possible, by insisting on the threat of a penalty, to abide by agreements made to achieve the desired result. It is best if this is already part of the contract. It may be necessary to ask the management for assistance.

But corrective actions must not be taken too soon!

6 COMMUNICATION

A big stumbling block in the project is when communication in the team is bad, the team members are deliberately or unintentionally not or only late informed.

In order to avoid this, clear rules are set by the project manager. For example, how feedback is given.

In project meetings, attention is also paid to clear communication. Conflicts are resolved immediately.

Communication distinguishes between two areas.

- o **Verbal communication**
  The spoken word is only 1/7 of communication.

- o **Non-verbal communication**
  It is important to promote non-verbal communication in addition to verbal communication (the spoken word).
  In addition to body language, gestures, facial expressions, tonality, etc., they are also important.
  Important information can be read here. Non-verbal communication applies in case of doubt.

Leading project meetings

Project meetings are a very important part of the project manager's work. They take up to 20% of his time
related to the respective project.

However, negotiations are also an important tool for
achieving the success of a project, for example by
identifying problems early on.

It is very important to work together during meetings if
they are to lead to a positive goal. It must be ensured
that the parties have equal rights and that everyone is
aware of this.

So no party has the power to force a result. That's what
negotiating skills are all about.

Rules of communication for meetings

- **The preparation**
  The meeting leader must know the goal of a
  meeting or negotiation exactly and must also in-
  form the participants exactly.
  An important part of the definition of goals is the
  consideration of what to do if no agreement can
  be reached according to the project manager's
  ideas or project requirements.
  The influence on the outcome of such negotiations is based on attractive alternatives, which

can then be presented.

The easier it is for the project manager to break off or postpone negotiations, the stronger his negotiating position becomes.

All participants in a meeting, including the project manager, are obliged to do their "homework", i.e. no participant may come into the meeting unprepared. All those involved should (can) take the time they need to prepare, even if they have to ask for postponement.

o **Minimization of perception differences**

The image of the event that a person makes for himself is based on his "history" and his experience.

This can differ significantly from that of the negotiating partner. Therefore, you must never assume that you know the other person's point of view.

Only through questioning can ambiguities be eliminated and agreement reached.

Therefore, it is very important to clearly define one thing so that the other person can confirm or correct the image.

Question techniques, for example from the target survey can help.

Through these questions, the situation can be

precisely grasped. Through close observation, you can see what is happening in front of you and then take the right measures.

- o **Listening**

  Active, attentive listening is an obligation for effective negotiation.

  The meeting leader must let the others have their say. When he talks about 50% of the time himself, he clearly doesn't listen enough.

  This also includes respecting silent breaks.

  The new impressions must first be processed before it makes sense to continue. No one should be tempted to fill in these creative pauses with speeches.

- o **Taking notes**

  It is necessary to write down what is discussed and what is decided.

  It does not make sense to rely on memory alone in the case of high stress. The agreements reached are therefore summarised in a memorandum.

  In addition to the measures, this memorandum (protocol) must also specify the persons responsible and the deadlines. If the project manager cannot do this himself, he appoints a participant before the meeting.

○ **Contributing creativity**

A premature end of the session or nagging criticism of statements made by employees dampens the spontaneity of the participants.

Time should always be allowed for alternative or unusual solutions to the problem.

During such a discussion all ideas can be presented value-free, i.e. without criticism. All negotiations, but also all further cooperation, can benefit from such a creative approach.

○ **Support the other party**

Good negotiating partners recognize that the problem of the other party is also their own problem.

To this end, they put themselves in the other party's shoes and work together with it to find a solution acceptable to all. After all, an agreement only holds if it is supported by all.

○ **Make compromises**

"Give something for nothing" should be avoided. At least one promise of goodwill or future repayment must remain for giving.

Even if you don't value the promise you have

made as much as the other person, there is always something that is more important to you than the other side.

o **Make apologies quickly**

If the choice of the word is misused during the negotiation, an apology is the quickest and safest way to reduce other people's negative feelings.

This is not only necessary when it comes to a personal apology. An apology about the current, muddled situation can also be effective.

Thus, hostile remarks should not contribute to a bad discussion climate.

Hostility directs the discussion from the essential to a level of self-defence at which one wants to harm the other.

o **Avoid ultimatums**

An ultimatum always requires the other side to either give up or fight to the bitter end.

Neither of these results will be conducive to future positive cooperation in a project.

Therefore, it should be avoided to drive someone into a corner. This happens, for example, when you only offer the party alternatives that are unacceptable to the other side.

A back door should always be left open for the other person.

o **Set realistic deadlines**

Many negotiations drag on too long because there is no deadline pressure (schedule for the meeting).

A deadline requires both sides to use their time effectively.

A realistic deadline pressure encourages both sides to make concessions and compromises. However, deadlines must be avoided that are unrealistic and cannot be met.

7 RULES FOR THE CREATIVE PROJECT TEAM

- o **Project team**

  Create an atmosphere of trust and collegiality
  "Promoting a "sense of togetherness" in the project
  Aim for open, comprehensive communication
  Identify and resolve conflicts as early as possible
  Recognizing and passing on good performance
  Team members show mutual respect and trust

- o **Conflicts and problems**

  Conflicts and problems are openly addressed and solved. Feelings such as anger are openly expressed and not suppressed
  All are equal partners and no one dominates

- o **Different opinions are perceived as a contribution to the solution of the problem**

  Only constructive criticism is practiced.
  Criticism only serves the progress of the project and is not personal. Different opinions contribute to solving the problem.

o **Defined roles and responsibilities**

   Every team member in high performance teams understands what they have to do or not do to support the success of the project.

o **Coordinating relationship of all team members**

   The individual work steps enable efficient coordination within the team.

o **A team strives for consensus**

   All team members abide by decisions made once.

o **Team meetings**

   All adhere to the "rules of the game": good preparation in front of meetings, completion of the tasks set by the deadline, punctuality Information about the project will be shared with everyone.

   The activities of all team members are known to all others. No team member carries out activities without prior consultation with the project manager.

   The tasks such as logging, monitoring of lists etc. are fairly distributed

- o **Record technical progress**

  Summarize recommendations for future development

  Summarising what has been learnt through cooperation

- o **Services**

  Write performance reports about the members of the project team

  Giving all employees feedback on their performance

- o **Project completion**

  Perform final inspection

  Write final report

  Leading a project review with the core employees

  Declare the project closed

8 TO DO LISTS AT THE END OF THE PROJECT

- o **Complete the last tasks in the project**

  The project result is checked for functionality.

  The operating instructions are written.

  The last plans are made.

  The project result is delivered to the client.

  The client's staff is trained in the use of the product or the equipment.

- o **Relocate project work**

  Employees are transferred to their department or assigned to a new project.

- o **Clean up resources at the end of the project**

  Excess equipment and material is taken to a new location or disposed of.

  Facilities and premises no longer required will be returned.

- o **Write performance reports about the members of the project team**

  The project manager records his feedback on the performance of the respective employees in writing and forwards it to the responsible head of department who is responsible for performance evaluation.

- o **Give all employees feedback on their performance**

  The project manager gives his feedback on the performance directly to the respective employees.

- o **Perform final inspection**

  A final inspection is best carried out with a checklist in which all essential points are listed.

- o **Write final report**

  The final report will be written. It is part of the project documentation.

- o **Leading a project review with the core employees**

  To look back on the project, the most important employees should exchange their experiences with each other and record them for other projects.

  The problems that have arisen and the solutions used are summarized.

  Positive, negative experiences and what has been learned through cooperation are recorded in writing.

  Record technical progress.

Recommendations for future development are made.

- o **Declare the project closed**

  It sounds banal, but it has a great effect to declare a project completed: The subconscious can check it off and let go.

The points help to benefit from the experience gained in the completed project for future projects.

The more carefully this happens, the better it is for the future project teams.

LITERATURE

Alphanodes: Meindl, C., Scrum-Rollen – Der Product Owner, Internet Stand: 2015-12-01, (https://alphanodes.com/de/product-owner)

Arden, Paul; Egal, was du denkst, denk das Gegenteil, Bastei Lübbe (Lübbe Ehrenwirth); Auflage: 5 (2011)

Berckhan, Barbara; Die etwas andere Art, sich durchzusetzen-Selbstbehauptungstraining für Frauen; dtv, 2003

Birkenbihl, Vera; Trotzdem lehren (MVG Verlag bei Redline), Moderne Verlagsges. Mvg, 2013

Birker, Gabriele; Birker, Klaus; von Pepels, Werner (Hrsg.); Teamentwicklung und Konfliktmanagement, Effizienzsteigerung durch Kooperation, Berlin, 1. Auflage, 2001

Bozyazi, E.: Design Thinking im Projektmanagement, Lehrmaterial Hochschule, Mannheim, Oktober 2014

Cameron, Julia; Von der Kunst des Schreibens und der spielerischen Freude, Worte fließen zu lassen, Droemer Knaur, (2003)

Covey, Stephen R. /Merrill, A. Roger /Merrill, Rebecca R; First Things First, Fireside by Simon & Schuster, New York, 1995

Covey, Stephen R. /Merrill, A. Roger /Merrill, Rebecca R.; Der Weg zum Wesentlichen: Zeitmanagement der vierten Generation, Campus Verlag, Frankfurt /M., New York, 2014

Csikszentmihalyi, Mihaly; Kreativität: Wie Sie das Unmögliche schaffen und Ihre Grenzen überwinden, Klett-Cotta, 2015

Denkmotor: Was ist Design Thinking?, Internet Stand 2015-12-05, Video, (http://www.denkmotor.com/angebot/kreativitat-und-innovation/seminardesign-thinking/)

Dörner, Dietrich; Die Logik des Mißlingens-Strategisches Denken in komplexen Situationen, rororo, 2015

Doran, G. T.; There's a S.M.A.R.T. way to write management's goals and objectives. Management Review, Volume 70, Issue 11(AMA FORUM), pp. 35-36, 1981

Dulabaum, Nina L.; Mediation: das ABC, die Kunst, in Konflikten erfolgreich zu vermitteln, Weinheim, 4. neu ausgestattete Auflage, 2009

Ernst, Heiko; Können wir unserem Bauchgefühl vertrauen?, Psychologie heute, 03/2003, S.20 ff

Eyer, Eckhard (Hrsg.); Report Wirtschaftsmediation, Krisen meistern durch professionelles Konflikt-Management, Düsseldorf, 2004

Francis C.; Young D.; Tuckman & Jensen, 1977

Gelb, M. J.; Das Leonardo-Prinzip, Econ Tb, 2001

Glasl, Friedrich; Konfliktmanagement, ein Handbuch für Führungskräfte, Beraterinnen und Berater, Bern, 7. Auflage, 2002

Goldberg, Natalie; Schreiben in Cafés, Natalie Goldberg, Autorenhaus, Auflage: 2. Auflage, 2009

Haeske, Udo; Team-und Konfliktmanagement, Teams erfolgreich leiten, Konflikte konstruktiv lösen, Berlin, 2013

Hanisch, Christian; Ausbildung zum Aufstellungsleiter „Systemischer Coach ICI /Systemischer Aufstellungs-leiter ICI", European Business Ecademy, Seven Mirrors Consulting GmbH, 2011

Hansel, J.; /Lomnitz, G.; Projektleiter-Praxis- Erfolgreiche Projektabwicklung durch verbesserte Kommunikation und Kooperation, Springer-Verlag, Berlin, 1987

Haynes, Marion E; Projekt-Management-Von der Idee bis zur Umsetzung -, Wirtschaftsverlag Carl Ueberreuter, Wien, 2003

Hofstetter, H.; Der Faktor Mensch im Projekt, In: Schelle, H., Reschke, H., Schnopp, R., Schub A. (Hrsg.): Loseblattsammlung "Projekte erfolgreich managen", Veröffentlichungen des Verbands Deutscher Maschinen- und Anlagenbau e. V., Köln, 1998

http://www.psy.lmu.de/soz/studium/downloads_folien/ws_09_10/muf_09_10/muf_schattke_0910.pdf

https://wissensarbeiter.wordpress.com/2015/06/30/produkte-und-projekte-was-sind-die-unterschiede/

Kotter, John; Rathgeber, Holger; Stadler, Harald ; Das Pinguin Prinzip-Wie Veränderung zum Erfolg führt-, Droemer, 2011

Kriegisch, A., Scrum - Auf einer Seite erklärt, Internet Stand: 2015-12-01, (http://scrum-master.de/Was_ist_Scrum/Scrum_auf_einer_Seite_erklaert)

Kriegisch, A.: Scrum-Rollen–Product Owner, Scrum Master, Internet Stand: 2015-12-01, (http://scrum-master.de/Scrum-Rollen/Scrum-Rollen_Product_Owner)

Küstenmacher, Werner Tiki; Simplify your life-einfacher und glücklicher Leben-; Knaur TB, 2011

Litke, Hans-D.,; Projektmanagement: Methoden, Techniken, Verhaltensweisen, Carl Hanser Verlag, München /Wien, 2015

Lohmann, Friedrich; Konflikte lösen mit NLP, Techniken für Schlichungs-und Vermittlungsgespräche, Paarberatung und Mediation, nach Virginia Satir, John Grinder und Thies Stahl, Paderborn, 2003

Lundin e. a.; Fish-Ein ungewöhnliches Motivationsbuch-; Ueberreuther Wirtschaft; 2015

Massow, Martin; Gute Arbeit braucht Zeit-Entdeckung der kreativen Langsamkeit; Heyne, 1999

Mehrmann, Elisabeth; Wirtz, Thomas; Effizientes Projektmanagement, ECOn Taschenbuch Verlag, Düsseldorf, 2. Auflage, 2002

Meise, Sylvie; Hör doch mal zu!, Psychologie heute, 07/2003, S.46 ff

Michel, Reiner M.; Projektcontrolling und Reporting, Sauer-Verlag, Heidelberg, 1989

Modler, Peter; Das Arroganz-Prinzip: So haben Frauen mehr Erfolg im Beruf, Krüger, Frankfurt, 2012

Motamedi, Susanne; Körpersprache – schwere Sprache, Psychologie heute, 10/1996, S.52 ff

Mühlisch, Sabine; Das Prinzip KörperSprache im Unternehmen: Inspirationen für eine lebendige Arbeitsgestaltung, Junfermann, 2014

Naumann, Frank; Diplomatie: Der sanfte Weg zum Sieg, Psychologie heute, 11/2003, S 64 ff

Nöllke, Matthias; Schlagfertigkeit; Haufe; 2015

Orthwein, M.; Meßmer D., Agiles Projektmanagement – Projektentwicklung mit Scrum, Kanban & Co., Techdivison, Internet Stand 2015-12-06; (https://www.techdivision.com/_Resources/Persistent/a 90c984a454ba0b8478694b83f7a8822514b8fc8/Agiles-PM-Whitepaper0502.pdf)

Pantalon, M.V.; Nicht warten- Starten – Das 7-Minuten-Programm zur Motivation, dtv premium, 2012

Rahn-Huber, Ulla; Der Vampir neben dir; Kreuz, 2002

Schelle, Heinz; Projekte zum Erfolg führen, dtv Verlagsgesellschaft; Auflage: 7, 2014

Schmidt-Tanger, Martina; Kreische, Jörn; NLP-Modelle-Fluff & Facts, VAK Verlag für angewandte Kinesiologie GmbH, Freiburg im Breisgau, 2005

Schulz von Thun, Friedemann; Miteinander reden, rororo Rowohlt-Verlag, Reinbek bei Hamburg, 2010

Schwarz, Gerhard; Konfliktmanagement, Konflikte erkennen, analysieren, lösen, Wiesbaden, 2013

Seiwert, Lothar J.; Das 1 x 1 des Zeitmanagements, Knaur Ratgeber, 2014

Seiwert, Lothar J.; Wenn Du es eilig hast, gehe langsam, campus, 2012

Stoss, Karl; Botschen, Günther; Management der Strategischen Geschäftseinheiten, Verlag Gabler, Wiesbaden, 1994

Süß, Gerda; Eschlbeck, Dieter; Der Projektmanagement-Kompass, So steuern Sie Projekte kompetent und erfolgreich, Braunschweig, 1. Auflage, 2002

Thomas, Carmen; ModerationsAkademie für Medien und Wirtschaft, Engelskirchen

Wazlawick, Paul; Anleitung zum Unglücklichsein, Piper, 2009

Wazlawick, Paul; Beavin, Janet H.; Jackson, Don D.; Menschliche Kommunikation. Formen, Störungen, Paradoxien, Huber Hans, 2011, Die 5 Axiome der Kommunikationstheorie von Paul Watzlawick

Weidner, Christopher A.; Feng Shui gegen das Chaos auf dem Schreibtisch, rororo, 2004

Weyer, Simone; Konfliktmanagement im Projekt, Diplomarbeit FH Bochum, Fachbereich Wirtschaft, 2004

Will, Franz; Teamkonflikte erkennen und lösen: Zwischen Emotionen und Sachzwängen, Beltz; 2012

Wolf, Axel; Macht: Wer dominiert wen?, Psychologie heute, 01/1999, S 20 ff

Young; A Technique for Producing Ideas, Mcgraw-Hill Professional, 2003

FEEDBACK

Thank you for a positive review

If you liked the book, please send me a positive review on Amazon Kindle.

Comments, questions or criticism

Here you can send me your comments, questions or criticism about the book "7 Boyant Checklists for Your Project".

In the Google form you can write me directly and you can book a Strategy Session here.

Sachwörterverzeichnis

ANHANG: LISTE DER LINKS

Free Strategy Session http://bit.ly/2FBysxb

Contact https://www.kisp.de/kontakt

Blog „Selbstführung & Produktivität"
https://www.kisp.de/blog

Google-Formular https://forms.gle/
ay1MJThQo9vV6bL89

Bonus material https://www.kisp.de/qcyr

ABOUT THE AUTHOR

Prof. Dr. Annette Kunow taught mechatronics and mechanical engineering at the Bochum University of Applied Sciences for 32 years after several years in industry.

Prof. Dr. Annette Kunow supports people to help themselves and structure them better. In addition companies that want more efficiency in their project management to bring me the money. Last but not least, she stands for startups in the area of Engineering as a Science Angel.

Her passion is to consistently advance things.

Annette Kunow is the author of several books.

Technische Mechanik Statik

Die Technische Mechanik ist eine Kernkompetenz eines jeden Ingenieurs. Ohne diese Kenntnisse können die physikalischen Eigenschaften von Systemen nicht erfasst werden.

Was Sie in diesem Buch lernen werden

1. Mathematische Grundlagen
2. Arbeitsbegriff der Statik
3. Gleichgewicht
4. Schnitt- und Reaktionskräfte
5. Haftung und Reibung
6. Raumstatik

Technische Mechanik Statik Übungen

Die Technische Mechanik ist eine Kernkompetenz eines jeden Ingenieurs. Ohne diese Kenntnisse können die physikalischen Eigenschaften von Systemen nicht erfasst werden.

Vollständig und mit möglichen Lösungsvarianten gelöste Übungsaufgaben

Was Sie in diesem Buch lernen werden

7. Mathematische Grundlagen
8. Arbeitsbegriff der Statik
9. Gleichgewicht
10. Schnitt- und Reaktionskräfte
11. Haftung und Reibung
12. Raumstatik

Technische Mechanik Elastostatik

Die Technische Mechanik ist eine Kernkompetenz eines jeden Ingenieurs. Ohne diese Kenntnisse können die physikalischen Eigenschaften von Systemen nicht erfasst werden.

Was Sie in diesem Buch lernen werden

1. Deformationen
2. Elastizitätsgesetz
3. Spannungen
4. Spannungszustände
5. Statische Bestimmtheit
6. Arbeitsbegriff der Elastostatik

Technische Mechanik Elastostatik Übungen

Die Technische Mechanik ist eine Kernkompetenz eines jeden Ingenieurs. Ohne diese Kenntnisse können die physikalischen Eigenschaften von Systemen nicht erfasst werden.

Vollständig und mit möglichen Lösungsvarianten gelöste Übungsaufgaben

Was Sie in diesem Buch lernen werden

7. Deformationen
8. Elastizitätsgesetz
9. Spannungen
10. Spannungszustände
11. Statische Bestimmtheit
12. Arbeitsbegriff der Elastostatik

Technische Mechanik Dynamik

Die Technische Mechanik ist eine Kernkompetenz eines jeden Ingenieurs. Ohne diese Kenntnisse können die physikalischen Eigenschaften von Systemen nicht erfasst werden.

Was Sie in diesem Buch lernen werden

- Kinematik
- Kinetik des Massenpunktes
- Kinetik des Massenpunktsystems
- Kinetik des Starrkörpers
- Ebene Bewegung
- Schwingungen

Technische Mechanik Dynamik Übungen

Die Technische Mechanik ist eine Kernkompetenz eines jeden Ingenieurs. Ohne diese Kenntnisse können die physikalischen Eigenschaften von Systemen nicht erfasst werden.

Vollständig und mit möglichen Lösungsvarianten gelöste Übungsaufgaben

Was Sie in diesem Buch lernen werden

- Kinematik
- Kinetik des Massenpunktes
- Kinetik des Massenpunktsystems
- Kinetik des Starrkörpers
- Ebene Bewegung
- Schwingungen

Projektmanagement und Business Coaching

Grundlagen des agilen Projektmanagements mit Methoden des Systemischen Coachings

Projektkompetenz ist heute die Kernkompetenz für jeden Berufstätigen. Ohne die Strukturierung durch das Projektmanagement sind Abläufe in Unternehmen nicht mehr zu bewältigen.

Was Sie in diesem Buch lernen werden

- Strukturierte Pläne
- Optimale Nutzung der Ressourcen
- Klar bewertbare Projektziele
- Angepasste Informationssysteme
- Führung des Teams
- Strategische Projektziele

Project Management and Business Coaching

Basics of Agile Project Management With Methods of Systemic Coaching

Project competence is today the core competence for every professional. Without structuring through project management, processes in companies can no longer be mastered.

What You Will Learn in this Book

- Structured plans
- Optimal use of resources
- Clearly assessable project objectives
- Adapted information systems
- Leadership of the team
- Strategic project goals

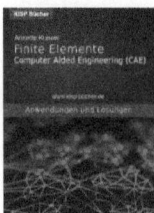

Finite-Elemente Methode / Computer Aided Engineering (CAE)

Theoretische Grundlagen und Lösungen

CAE ist heute in den Konstruktions- und Entwicklungsbereichen der Industrie nicht mehr wegzudenken. Die heute übliche automatische Vernetzung kann ohne das Grundlagenwissen zu gravierenden Fehlern führen.

Was Sie in diesem Buch lernen werden

- Grundbegriffe und Gesamtsteifigkeit
- Flächen- und Volumenelemente
- Vernetzungsregeln
- Versuche
- Dynamische Berechnungen
- Nichtlinearität

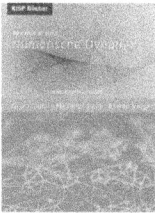

Numerische Dynamik

Grundlagen-Modellbildung-Anwendungen

Die Numerische Dynamik ist ein bedeutender Bestandteil im Engineering. Sie vermittelt die physikalischen Zusammenhänge, um Konstruktionen unter bewegten Belastungen zu dimensionieren.

Was Sie in diesem Buch lernen werden

- Grundlagen der Dynamik/ Kinetik
- Prinzip der dynamischen Berechnung
  - Einmassenschwinger
  - System mit zwei Freiheitsgraden
  - Mehrmassensystem
- Berechnung für das Kontinuum
- Ausführliche Beispiele und Übungen, incl. Eingaben in die Programme (EXCEL, MATLAB)

Numerische Dynamik Übungen

Grundlagen-Modellbildung-Anwendungen

Die Numerische Dynamik ist ein bedeutender Bestandteil im Engineering. Sie vermittelt die physikalischen Zusammenhänge, um Konstruktionen unter bewegten Belastungen zu dimensionieren.

Was Sie in diesem Buch lernen werden

- Grundlagen der Dynamik/Kinetik
- Prinzip der dynamischen Berechnung
  - Einmassenschwinger
  - System mit zwei Freiheitsgraden
  - Mehrmassensystem
- Berechnung für das Kontinuum
- Ausführliche Beispiele und Übungen, incl. Eingaben in die Programme (EXCEL, MATLAB)

24 raketenstarke Produktivitäts-Stategien

So bringen Sie Ihren ganz persönlichen Business-Alltag zum Abheben

In diesem Booklet finden Sie eine Zusammenfassung der wichtigsten Produktivitäts-Strategien.

Was Sie in diesem Buch lernen werden

- o Zusammenfassung der wichtigsten Produktivitäts-Strategien

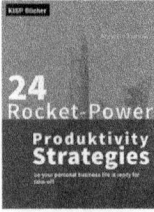

24 Rocket Power Productivity Strategies

How to make your very personal business day stand out

In this booklet you will find a summary of the most important productivity strategies.

What You Will Learn in this Book

o   Summary of the most important productivity strategies

7 auftriebsstarke Listen für Ihr Projekt

So bringen Sie Ihr nächstes Projekt in Sekunden-schnelle nach oben

Ich zeige Ihnen hier meine 7 wichtigsten Checklis-ten zur Planung und erfolgreichen Durchführung eines Projektes.

Und Sie bekommen eine umfangreiche Beschreibung zu den ein-zelnen Schritten, um es gut umsetzen zu können.

Was Sie in diesem Buch lernen werden

o   Checklisten zur Planung und erfolgreichen Durchführung eines Projektes mit Beschreibunmhg

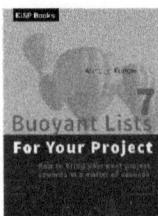

7 Buoyant Lists For Your Project

How to bring your next project upwards in a matter of seconds

Here I show you my 7 most important checklists for planning and successful implementation of a project.

And you will receive a comprehensive description of the individual steps in order to be able to implement it well.

What You Will Learn in this Book

- o Checklists for planning and successful implementation of a project

www.ingramcontent.com/pod-product-compliance
Lightning Source LLC
Chambersburg PA
CBHW022059210326
41520CB00046B/773